Ancient Athenian Building Methods

AMERICAN SCHOOL
OF CLASSICAL STUDIES
AT ATHENS
1984

1. The Agora, showing at left the Hephaisteion and in the center the Stoa of Attalos. In the background are the two mountains which provided Athens with high quality marbles: Mount Pentelikon to the left of Lykabettos hill and Mount Hymettos to the right.

GREEK ARCHITECTURE, with its strong influence on later building styles, remains one of the great legacies passed down to us from antiquity. No visitor to Athens can fail to be moved by the grandeur and harmony of the classical monuments which crown the Akropolis. Equally impressive is the technical skill which went into the construction of Greek buildings.

The Agora, the heart of ancient Athens in all respects, is an ideal place in which to study Greek building techniques. With its imposing marble temples, large, functional public buildings, and modest private houses, the Agora affords an unparalleled range of building types (1, 26). In addition, the long history of the area provides an opportunity to study the development of Athenian building methods over a span of more than a thousand years, illustrating how these techniques evolved in the course of time. Finally, the reconstruction of the Stoa of Attalos in the 1950's provided unique insights into ancient Greek building practices. The evidence from the buildings themselves may be supplemented by numerous inscriptions: building accounts which include specifications of design and material as well as amounts of money actually paid for individual tasks as the buildings took form.

FINANCING AND LABOR

The first question concerning an ancient building might be, who paid for it? In Athens, the state provided funds for many of the large public buildings. If a temple were to be constructed, it would usually be paid for out of the sacred treasury; other buildings were financed with general funds or the spoils of a successful military campaign. In addition, many rich Athenians are known to have contributed important monuments to the city, and in later times foreign kings and Roman emperors were eager to honor Athens and aggrandize themselves by erecting a large public building in the famed seat of culture and education (2). King Attalos II of Pergamon, for instance, who had studied as a youth under the philosopher Karneades in Athens, dedicated the large stoa which borders the east side of the square (1).

Once the decision to build had been made, the labor force was assembled. It would be headed by the architect (ἀρχιτέκτων), the man responsible both for the design and for the supervision of actual construction. Buildings were carefully planned before work began, but it is clear from the remains that extensive modifications were sometimes made during the course of erection.

Building accounts of the Erechtheion on the Akropolis have survived in part, and from them it is evident that the work force was a varied one. Free Athenian citizens, resident aliens, and slaves all worked together on the building, side by side, and all were paid the same daily wage of one drachma, three times the minimum subsistence rate of two obols paid to recipients of public welfare. The number of workers involved must have been very large since many skills were required. The Erechtheion accounts record payments to more than 110 different workmen: masons, carpenters, sculptors, painters, and laborers. The modern reconstruction of the Stoa of Attalos needed 198 men.

The time required to finish an ancient building varied tremendously, depending, as today, on such factors as finances, politics, and wars. Except for final decoration, the Parthenon was built in a mere ten years, the unfinished Propylaia in five; at the other end of the scale, construction of the temple of Olympian Zeus was begun early in the 2nd century B.C. but only completed some 300 years later. The reconstruction of the Stoa of Attalos, which combined ancient and modern techniques, took four years.

2. Dedicatory inscription of the Library of Pantainos (*ca.* A.D. 100), carved on the lintel of the main doorway: "To Athena Polias and to the Emperor Caesar Augustus Nerva Trajan Germanicus and to the city of the Athenians the priest of the wisdom-loving Muses, T. Flavius Pantainos, the son of Flavius Menander the head of the School, gave the outer stoas, the peristyle, the library with the books, and all the furnishings within them, from his own resources, together with his children Flavius Menander and Flavia Secundilla."

3. The Hephaisteion, on the hill of Kolonos Agoraios, facing east over the Agora square. Immediately below the temple the hillside was cut back to make a terrace for the new Bouleuterion at the end of the 5th century B.C.

PREPARATION OF THE SITE

Choosing a site for an ancient building was an important part of the initial work. Temples by preference were placed at or near the top of a hill and usually faced east (3). Stoas, the long covered colonnades in which so many activities of Greek public life took place, often faced south to take advantage of the low sun in winter and to shut off the cold north winds.

Once the site was chosen, the ground had to be prepared. Since little of the terrain of Greece is naturally flat, leveling was often required, either by artificial terracing supported by heavy retaining walls or by cutting back into the hillside. The Bouleuterion (senate house) was set in a flat area created by quarrying out the bedrock at the base of the hill called Kolonos Agoraios (3), and both the Middle Stoa and the Stoa of Attalos had long raised terraces in front, supported by massive retaining walls (26). If necessary, provisions for drainage were made. The entire area of the Agora was drained by a system of large stone channels now known as the "Great Drain" (4), which carried off storm water from the area as well as disposing of sewage from the numerous buildings along its course.

4. The "Great Drain" (late 6th or early 5th century B.C.), built in the polygonal style of masonry. In this stretch, only one side is shown.

4

5. Terracotta pipeline, late 6th century B.C. The covered holes were for setting the line and to provide access for cleaning and repairs.

WATER SUPPLY

A supply of water was of major importance for many buildings and could be obtained by various methods. Round terracotta pipelines brought water to the city from distant springs to feed public fountain houses as early as the 6th century B.C. Short sections of pipe were fitted together with elaborate collars at the joints (5). Until the 3rd century B.C., all such aqueducts were simple gravity systems; thereafter pressure lines were used occasionally. Private establishments had their own water sources.

From as early as 3000 B.C. wells, plain, unlined shafts which tapped the underground water table, were sunk into the bedrock that underlies the city. From the 3rd century B.C. on, the sides of such shafts were usually reinforced to prevent the collapse of the soft bedrock. Cylindrical drums of tile lined the well (6), each drum made up of three or four segments clamped together and then set one on top of another. Such linings allowed the Athenians to carry wells as deep as 30 meters. Another method of supply, developed in the 4th century B.C., was provided by the bottle-shaped cistern, a waterproofed chamber cut in bedrock and designed to hold rain-water draining off the roof of a building. Although cement and mortar were not used for structural purposes in Greek architecture, good water-proof cement was employed at least as early as the 5th century B.C.

6. Tile-lined well of the 3rd century B.C. Letters were incised on the upper edges of the segments to ensure proper assembly of each drum.

7. Cross section of a bottle-shaped cistern (3rd century B.C.) cut into bedrock and lined with waterproof cement.

BUILDING MATERIALS

The Athenians employed a variety of building materials. Walls composed of mud brick or field stones, laid either dry or in clay and reinforced with wood timbers, appeared at a very early period and were common even in classical times for modest buildings. As early as the 7th century B.C. the Greeks cut blocks of an easily worked limestone (poros) for monumental buildings, and by the 6th century B.C. fine white marble from the islands of Naxos and Paros was being imported to Athens for both building blocks and sculpture. In the early 5th century the Athenians turned to sources closer to hand and began exploiting the rich quarries of excellent white marble on Mount Pentelikon, only 11 miles northeast of the city (1). This new material became a mainstay of Athenian public and religious buildings for the next 750 years and was heavily exported as well. The island marbles

8. Steps of the Metroon, 2nd century B.C. The Ionic column base is of white Pentelic marble, the steps of blue Hymettian marble, and the foundations of hard, gray limestone. The step block below the column base displays the usual treatment of a joint surface (*anathyrosis*).

9. Conglomerate foundations of the temple of Apollo Patroos with limestone insets (4th century B.C.). The conglomerate was so coarse that the insets were needed to hold the dowels that anchored the lowest step to these foundations.

10. Mud-brick walls resting on stone orthostates, with a floor of packed clay. South Stoa I, 430–420 B.C.

were reserved for special purposes such as architectural sculpture and decoration. Late in the 5th century the Athenians began to quarry on Mount Hymettos, just southeast of the city (1), for its blue or blue-gray marble (8). To these fine marbles the Athenians added various limestones which range considerably in color, texture, and hardness and which were quarried from near-by locations: the Akropolis, Piraeus, Eleusis, and Aigina. Breccia and conglomerate (9), extremely irregular in texture and gray or reddish in

11. Roman brickwork, early 5th century after Christ, southeast of the Stoa of Attalos. Baked brick construction began as early as the 1st century after Christ and continued through Byzantine times.

7

12. Wall revetted with nine different kinds of marble. Villa south of the Agora, 4th century after Christ.

color, appeared in the late 5th century, largely in foundations since they are too coarse to be dressed to a smooth surface (9, 16). Walls above ground made of these rough materials were stuccoed. Private houses and inexpensive public buildings continued to be built of unbaked mud brick set on stone socles or on orthostates (vertically set blocks) (10). Properly plastered, mud brick will last for centuries and is much cheaper and easier to build with than large, squared blocks of stone which must be quarried, transported, shaped, and placed in position, all with great effort. The introduction of oven-baked brick, mortar, and concrete in wall construction dates only from Roman times in the 1st century after Christ (11, 12). In most instances, the walls of this later period were covered with a thin revetment of marble to mask the brick or concrete behind, although only rarely has such revetment survived in place (12).

QUARRYING AND TRANSPORT

Building blocks were quarried according to the specifications of the architect, who stipulated the material, number, and size of the blocks required. Measurements were given in foot lengths, subdivided into 16 dactyls (fingers). The foot lengths varied between buildings, the two most common having modern equivalents of 0.294 and 0.327 meters. Ancient quarrying was slow. The length and width of the block were measured out on the surface of the rock, and continuous channels were cut down vertically with hammer and chisel until the desired height was obtained. Notches were then chiseled under the bottom of the block, wooden wedges were inserted and soaked in water, and as the wedges expanded they broke the block free. When quarried, a block was always several centimeters larger all around than the finished piece, the extra stone to serve as a protective surface during transportation. Small blocks could be carried on carts, but many of those required for Athenian buildings were huge, often weighing many tons. Special innovations were needed to move such enormous stones, and we know of great wooden wheels fitted around individual blocks which were then pulled along the rough roads by teams of draft animals. A building account at Eleusis lists payments for 33 teams of animals to haul a single column drum to the site; the 22-mile journey from the quarry took three days and cost 400 drachmas. It is clear that quarrying and transport of new blocks accounted for much of the cost of a building. For this reason material was often salvaged from abandoned structures. South Stoa II in the Agora was built in the 2nd century B.C. almost entirely of material taken from a large peristyle of the late 4th century B.C., and south of the Stoa of Attalos the fortification wall of the 3rd century after Christ is a pastiche of re-used blocks from a number of Agora buildings destroyed by the Herulians in A.D. 267 (13).

13. Late Roman fortification wall (*ca.* A.D. 267–280), built from an assortment of re-used material, following the destruction of the Agora in A.D. 267.

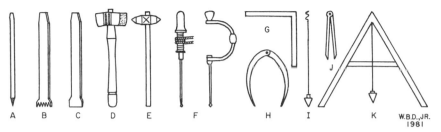

14. Ancient tools and measuring devices.

15. A selection of chisels used in the reconstruction of the Stoa of Attalos.

TOOLS

Various tools and instruments (14, 15) were devised for stoneworking. The cutting and finishing was done with saws, drills (F), and, primarily, with hammers and chisels (A–E). Chisels were of iron and came in a variety of sizes; their most common shapes were the point (A), the toothed chisel (B), and the flat or drove chisel (C). Tool marks characteristic of each are still visible on many ancient blocks. Marble might be polished with emery powder and leather. Different sorts of measuring devices insured the accuracy of workmanship: foot measures, squares (G), calipers (H), dividers (J), and an A-frame mason's level with plumb bob (K).

16. Conglomerate foundations of the Stoa of Attalos, *ca.* 150 B.C. The pier at the right, 17 courses deep, supported one of the columns of the interior colonnade.

WALLS

Once the site was prepared and the building materials assembled, actual construction began. Trenches were cut down, usually to bedrock, and the foundations laid. Generally, the hidden foundations (8, 16) were of a poorer material than the exposed walls above, which could be erected in one of several masonry styles (17). Among the earliest was polygonal, constructed of large blocks cut with four or more sides of irregular lengths and closely fitted together (4). In addition to its pleasing appearance, this style of wall had exceptional strength, with no natural lines of weakness. The most common type of masonry, however, was ashlar: large squared blocks set in regular courses (18).

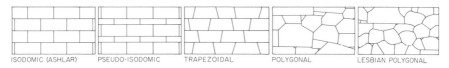

17. Athenian masonry styles.

18. Ashlar masonry, with lifting bosses left in place; back wall of the Stoa of Attalos.

Before the blocks were set in place, the rough, quarry surface had to be removed and the faces carefully worked down, often with a claw chisel first and then with a drove. The top and bottom resting surfaces of the block were completely smoothed. To safeguard the block during construction, the visible vertical faces retained a thin protective layer which was chiseled off only when the building was fully erected. The vertical joint surfaces were treated with *anathyrosis*, i.e., only a strip along the vertical edges and top was polished smooth, while the rest of the surface was recessed slightly and left rough picked (8). This treatment allowed a tight joint with a minimum of effort and expense since only the narrow strip of stone needed to make close contact with its neighbor; the roughened areas were hidden within the thickness of the wall.

There were a variety of means for lifting blocks into place (19). Most commonly employed were lifting bosses, rectangular protrusions of the original protective envelope which were retained on opposite faces of the block when the rest of the surface was cut smooth (18). Such bosses, located at the center of gravity, served as handles with either rope slings or

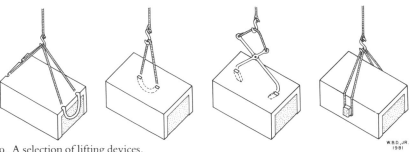

19. A selection of lifting devices.

20. A lewis, as used in the reconstruction of the Stoa of Attalos.

iron tongs. In the Classical period these bosses were always removed in the final stages of construction, and their presence, as on the Propylaia of the Akropolis, indicates that the building is unfinished. In Hellenistic times the bosses were often left for their decorative effect of light and shadow on an otherwise blank expanse of wall. As an alternative to lifting bosses, grooves might be cut into the blocks for lifting ropes which could easily be pulled out of these channels after placement. Another lifting device was the lewis, a set of flat and wedge-shaped iron bars with a ring on top for attaching a rope; these bars were set into trapezoidal cuttings on the tops of the blocks (20). All cuttings for these various lifting devices were placed in such a way as to be hidden in the finished wall.

Workers lifted the block onto the wall by means of a high tripod, multiple pulleys, and rope. Once in its proper course, the piece was moved closer to its correct location on wooden rollers. Crowbars were needed for the final positioning; to give them purchase, shallow indentations, or pry holes, were cut into the top of the course below (21, 23).

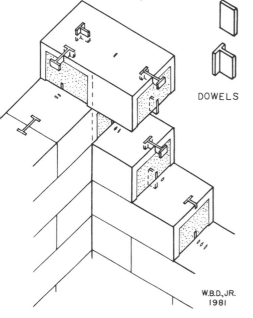

DOWELS

21. Drawing showing the use and position of clamps, dowels, and pry holes.

W.B.D., JR.
1981

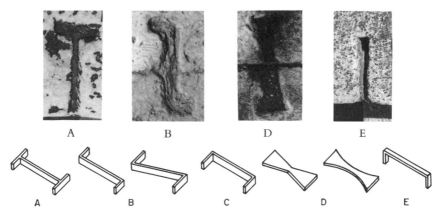

22. A selection of Athenian clamps: (A) double T, (B) Z-clamp, (C) U-clamp, (D) swallow-tail and (E) hook clamp.

As no mortar was used for construction during the Greek period, iron clamps and dowels (21) fastened the blocks together. Clamps at each end of a block held it horizontally to the adjoining block in the same course. The clamps came in a variety of forms, some shapes more common in certain periods than others (22, 23). Dowels were flat, rectangular bars, set into cuttings in the top and bottom surfaces of blocks, designed to join one course to another vertically (9, 21, 23). The cuttings for both clamps and dowels were larger than the actual metal fastenings; molten lead poured around the iron seated it firmly and prevented air and moisture from rusting and expanding the metal. Narrow channels allowed the molten lead access to hidden dowel cuttings.

23. Step block of the Temple of Ares, 5th century B.C., dismantled and moved to the Agora in the 1st century B.C. Large letters (ΠΕ) are masons' marks to ensure proper reassembling of the pieces. Other cuttings include T-clamps at either end, a dowel hole just below the Γ, and two pry holes.

24. Masons' marks at the east end of the Middle Stoa, 2nd century B.C. The upper pair of letters (YΔ, read sideways) refers to the course, and the lower pair (ΘΘ) refers to the position of the block within the course.

MASONS' MARKS

Occasionally, blocks bear masons' marks, usually in the form of letters. At the quarry a single letter might be inscribed to designate the job for which the block was intended or perhaps the mason or contractor responsible for the work. Most such marks, however, indicate where in the building the block was to be set (24). In many instances setting letters are not original with the building but were cut when an old building was dismantled, moved, and re-erected. The marks then ensured the replacement of the pieces in their correct positions. The Temple of Ares (23) and parts of the Temple of Athena at Sounion, the so-called stoa at Thorikos (25), and other buildings were dismantled, brought to the Agora, and re-erected in the Roman period, all these pieces being appropriately lettered.

25. Column drum from the so-called stoa at Thorikos, moved to Athens in the Roman period and re-used in a temple. The letter gamma (Γ) indicates the position of the drum within the column, the number of gammas indicates the column to which the drum belongs. The square cutting is for an *empolion* (30), and the joint surface of the drum has been treated with *anathyrosis*.

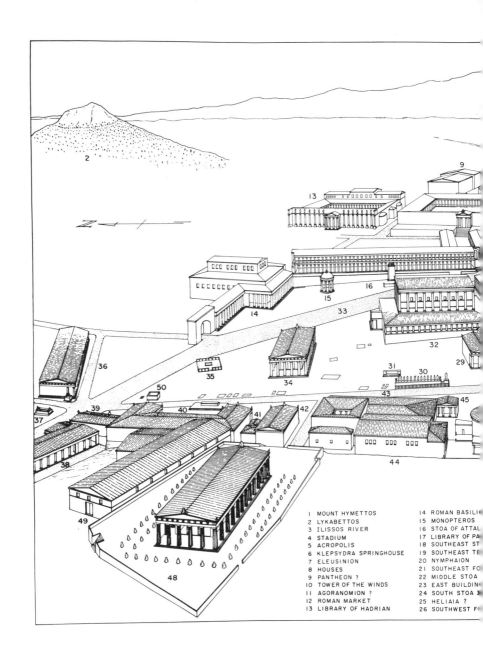

1	MOUNT HYMETTOS	14	ROMAN BASILI⬚
2	LYKABETTOS	15	MONOPTEROS
3	ILISSOS RIVER	16	STOA OF ATTAL⬚
4	STADIUM	17	LIBRARY OF PA⬚
5	ACROPOLIS	18	SOUTHEAST ST⬚
6	KLEPSYDRA SPRINGHOUSE	19	SOUTHEAST TE⬚
7	ELEUSINION	20	NYMPHAION
8	HOUSES	21	SOUTHEAST FO⬚
9	PANTHEON ?	22	MIDDLE STOA
10	TOWER OF THE WINDS	23	EAST BUILDIN⬚
11	AGORANOMION ?	24	SOUTH STOA ⬚
12	ROMAN MARKET	25	HELIAIA ?
13	LIBRARY OF HADRIAN	26	SOUTHWEST F⬚

ATHENIAN

AGORA

A.D. 150

TO THE BATHS

TO THE PRISON & TO PIRAEUS

27	TRIANGULAR SHRINE	40	STOA OF ZEUS ELEUTHERIOS
28	CIVIC OFFICES	41	TEMPLE OF ZEUS PHRATRIOS
29	SOUTHWEST TEMPLE		AND ATHENA PHRATRIA
30	EPONYMOUS HEROES	42	TEMPLE OF APOLLO PATROOS
31	ALTAR OF ZEUS AGORAIOS ?	43	METROON
32	ODEION	44	BOULEUTERION
33	PANATHENAIC WAY	45	PROPYLON TO BOULEUTERION
34	TEMPLE OF ARES	46	THOLOS
35	ALTAR OF THE 12 GODS	47	STRATEGEION ?
36	POIKILE STOA	48	HEPHAISTEION
37	ALTAR	49	ARSENAL ?
38	ROMAN STOAS	50	CROSS-ROAD SANCTUARY
39	ROYAL STOA		

W. B. DINSMOOR, JR.
1980

27. Doric capital from the so-called 'stoa' at Thorikos of the late 5th century B.C., re-used in the Agora in the early Roman period.

28. Ionic capital, from an unidentified building of the 5th century B.C. Note the traces of the original painted design between the volutes.

COLUMNS

The standard supporting member in most Greek buildings, particularly temples and stoas, was the round column. The most common type in Classical Athens was the simple Doric (27). The more slender Ionic (28) with its spiral capitals was favored in the Hellenistic period and the elaborate floral Corinthian (29), in Roman times, although both were known and used occasionally from the second half of the 5th century B.C. on. Except for the bases and capitals, the carving of columns was similar in all three types. The shafts were made up of a series of cylindrical drums. A large, square hole appears in the center of the top and bottom of each drum (25); the matching holes on adjoining drums were not for a dowel but rather for an *empolion*, one of a pair of wooden blocks into which a round wooden centering peg was placed in order to align the drums exactly on top of one another (30). The joint surfaces were finished and treated with *anathyrosis* (25), but a protective surface was left on the sides of each drum during erection of the column. With the exception of the top and bottom of the shaft, where the first few centimeters were finished before placement, the vertical channels (flutes) were carved after the column was standing to its full height. Carving the inceptive flutes, which served as guides for the

18

29. Corinthian capital, from the Odeion of Agrippa, late 1st century B.C.

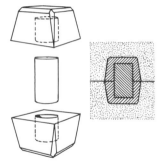

30. An *empolion*, designed to ensure the correct centering of the drum of a column (cf. Ill. 25).

final fluting of the shaft, would have been difficult once the column was in place. The cost of completely fluting a column of the Erectheion at the end of the 5th century B.C. was 350 drachmas (five men working about 70 days each). Starting in Hellenistic times, the lower third of the columns of stoas was often left unfluted (31). This procedure reduced the costs and also avoided the damage fluting was likely to receive from the large numbers of people usually accommodated in stoas. Columns were not invariably fluted, however, and in all periods there were buildings with smooth round shafts, both Doric and Ionic. When fluted, Doric columns had a sharp arris between flutes while Ionic columns employed a flat fillet.

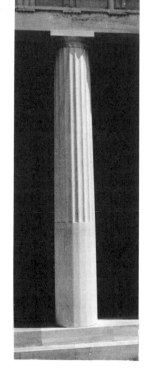

31. Replicated Doric column of the Stoa of Attalos, 2nd century B.C., unfluted at the bottom.

32. Doric entablature, southeast corner of the Hephaisteion, second half 5th century B.C. In this case the metope is decorated with a sculptured scene of Theseus fighting the Minotaur.

ENTABLATURE

The entablature over the columns of both Doric and Ionic buildings usually consisted of three members set one above the other: the architrave (epistyle), the frieze, and the cornice (geison). The basic elements of the two orders imitated wooden construction and remained unchanged for centuries. In the Doric order (32 and front and back covers) the architrave is a plain beam crowned by a horizontal band (taenia), under which appear short strips (regulae) with hanging pegs (guttae). The frieze, also capped with a horizontal band, is composed of alternating plain square panels (metopes) and rectangular blocks (triglyphs) divided into three vertical bands separated by V-grooves. The cornice overhangs the frieze and has a plain vertical face crowned with a molding; on its underside, separated by narrow channels (viae), it has rectangular flat panels (mutules) with three rows of hanging pegs.

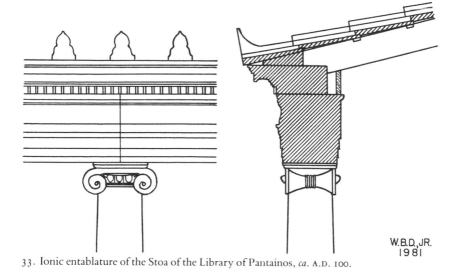

33. Ionic entablature of the Stoa of the Library of Pantainos, *ca.* A.D. 100.

W.B.D.,JR.
1981

In the Ionic and Corinthian orders (33) the architrave usually has three horizontal bands (fasciae), each projecting slightly beyond the one beneath, and a crowning molding. The frieze is a plain beam with a crowning molding. The cornice is similar to the Doric, but the underside is plain and gently curved. From the late 4th century B.C. on, a row of rectangular projections (dentils) was often added at the bottom of the cornice (34), a tradition from the original Ionic style of Asia Minor, where dentils appeared without a frieze, as they do in Athens, by exception, in the south porch of the Erechtheion.

34. Ionic entablature of the Southeast Stoa, Roman period, showing the use of dentils. Note also that frieze and architrave are carved from a single block.

Other changes in design may also be dated from the 4th century, when columns became more slender and entablatures more compressed. In the Doric order the number of metopes between columns was often increased from two to three, and in the Ionic order the frieze and epistyle of small-scale buildings were often carved on a single block for added strength (34).

35. Coffered ceiling, east porch of the Hephaisteion, 5th century B.C.

CEILINGS

In temple architecture, interior ceilings were made of wood; those between the outer colonnade and the cella were usually of limestone or marble. These exterior ceilings were normally formed of large stone slabs into which were cut coffers in imitation of wooden ceiling construction. These deep, square depressions were richly painted or carved or both (35–37). In addition to the decorative effect, the cutting of the coffers reduced the weight which had to be borne by the beams.

37. Watercolor of the painted design on a ceiling coffer from the Temple of Ares, 5th century B.C. The principal colors used are dark blue for the background, red and gold for the decoration.

36. Coffered ceiling, east porch of the Hephaisteion.

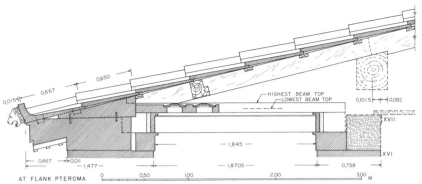

38. Cross section of the roofing system of the Hephaisteion, 5th century B.C.

ROOFING

The roofs of Athenian public buildings required massive wooden supporting beams, most of which had to be imported. Inscriptions refer to the importation of wood from Macedonia, Corinth, and the island of Karpathos, but the strongest timbers came from the cedar forests of Lebanon. The back of the marble tiles used on temples rested on strips of wood (battens) which lay across the rafters, and the front rested on the edge of the next tile down (38, 39). For ordinary roofs, terracotta (baked clay) tiles were set on a layer of clay and straw which rested on a wooden deck (39).

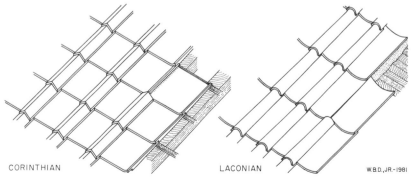

39. Corinthian and Laconian roofing systems. The Corinthian system shown is for marble roofs, the tiles resting on battens. The Laconian system illustrates the more common method employed by both systems for terracotta tiles.

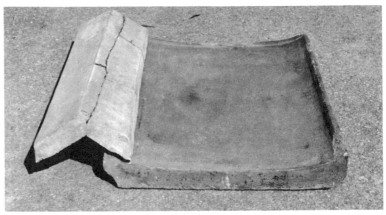

40. Corinthian pan and cover tiles, of terracotta.

41. Laconian pan and cover tiles, of terracotta.

There were two systems of pan and cover tiles for the roof: large, flat pan tiles with angular cover tiles, known as Corinthian, and curved pan and cover tiles, known as Laconian (40, 41). The pan tiles in either case were laid side by side with the bottom edge of each row overlapping the upper edge of the row below; cover tiles, also overlapping, were then set over the joints so that the entire roof was waterproof (39). The slope of most Greek roofs measured about one vertical unit for every four horizontal so that rainwater could not be blown under the overlapping joints.

42. Painted terracotta sima from South Stoa I, late 5th century B.C.

43. Restored drawing of the terracotta sima of the Middle Stoa, mid-2nd century B.C.

SIMAS

The row of eaves tiles along the edge of the roof, on top of the cornice, was ornately painted and assumed one of two forms, flat or turned up. If the outer edge turned up in order to form a gutter (sima), rainwater on the roof was thrown clear of the building by means of spouts, often in the form of lions' heads (38, 42, 43, and front cover).

44. Mold for the manufacture of a terracotta sima, Hellenistic period.

45. Cast taken from mold (44).

25

46. Marble antefix from the Stoa of Atta-los, mid-2nd century B.C.

47. Terracotta antefix, Roman period.

ANTEFIXES

If the eaves tiles were flat, no spouts were necessary. The ends of the lowest cover tiles were exposed in this case, however, and had to be given a decorative termination (antefix). Originally, antefixes took the form of a semicircular disk or the head of a gorgon or human; by the Classical period the most common form of antefix by far was the palmette (46–49). Occasionally, antefixes were used together with a sima, even though the ends of the cover tiles would have been hidden (43 and front cover).

48. Fragment of a mold for the manufacture of a terracotta antefix, Roman period.

49. Cast taken from mold (48).

50. Marble standard for roof tiles, Roman period.

Roofing systems eventually became so regular that the tiles came in standard sizes (50). Buildings of unusual type, however, required specially designed roof tiles. The Tholos, for instance, was covered with elaborate triangular and diamond-shaped tiles made specifically to roof the round form of the building (52). The manufacture of terracotta tiles was itself no mean feat; during the reconstruction of the Stoa of Attalos about one third of the roof tiles misfired and had to be discarded.

51 Fragment of a stamped roof tile from the Metroon, 2nd century B.C. "Consecrated to the Mother of the Gods. Dionysios and Ammonios." Many roof tiles like this were stamped with the makers' names.

52. Eaves tiles and antefix from the roof of the Tholos, first half of the 5th century B.C. The round form of the building required special tiles of unusual shape.

53. Threshold block, showing doorstop as well as cuttings at edges for pivots and at center for a lock. Roman villa, southwest of the Agora.

DOORS AND WINDOWS

Doors were generally of wood, occasionally sheathed in bronze; much of our information about them comes from vase paintings and from marble imitations in Macedonian tombs. In Athens, stone door frames and particularly the thresholds remain, revealing much about the missing doors. A single, large block of hard stone was chosen for a threshold (53). Part of the top surface was cut down to provide a ledge which acted as a doorstop. If, as was usual, the door closed against wooden jambs, rectangular cuttings at the ends of the threshold show their position. At the outer corners of the lower area in Illustration 53 are round or square cuttings within which the door pivots rested and rotated as the door opened and closed. Other cuttings in the center served to anchor bolts for locking the door.

The interiors of Greek buildings had little natural light, most of it coming through open doors. Houses were designed to face inwards onto an interior courtyard for privacy and security. Any exterior windows would have been very narrow and raised high above the ground, openings in the wall which in winter might be covered with an oiled cloth. The windows in the Stoa of Attalos are narrow vertical slots which widen on the interior.

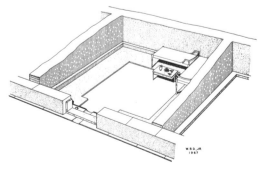

54. South Stoa I, dining room, late 5th century B.C. The wooden dining couches rested on a platform raised above the floor to protect them from water damage. The doors of most dining rooms were set off center, to accommodate these characteristic Greek couches.

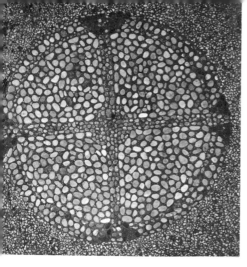

55. Mosaic floor of beach or river pebbles, from a house west of the Areopagus, 4th century B.C.

FLOORS

Many Athenian buildings, even some important ones like the Royal Stoa, had simple floors of packed clay. Temples and other public buildings were often provided with a more elegant floor of large, marble paving slabs or plain pebble mosaic. Dining rooms were frequently given special treatment in the form of a patterned mosaic floor which was watered down to enliven the colors and to provide air cooling by evaporation (54). These floors were made by setting pebbles or chips into a lime-mortar or cement bedding. Such mosaics are known as early as the 5th century B.C.; the earliest were of rounded beach pebbles in a limited number of colors (55). In later mosaics, special square-cut chips (tesserae) supplied a full range of colors (56). Often the design was a simple geometric pattern, although many mosaics included floral designs and scenes from both mythology and daily life.

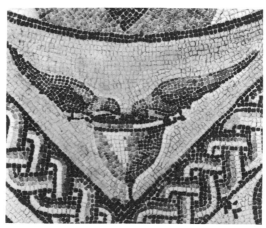

56. Watercolor of a mosaic floor of cut tesserae, from a house west of the Areopagus, 2nd century after Christ.

57. Anta (pier) capital of the 5th century B.C., showing the use of
paint and, at left, the profile of the moldings.

DECORATIVE ELEMENTS AND PAINTING

Decorative moldings regularly adorned certain parts of Greek buildings,
for example, the base and capital of an Ionic column (8, 28), the top of a wall
or pier (57), the underside and top of a cornice (front cover). The canons of
architecture were so rigid for the Doric and Ionic orders that a specific
molding was almost invariably associated with a given location on a building. Slight changes in the proportions of these decorative elements often
give an indication of their date.

58. Carved moldings on the altar of Zeus Agoraios (?), 4th century B.C. From the bottom up: a guilloche, a Lesbian leaf, a bead-and-reel.

Athenian buildings were richly painted as well, again according to a set pattern. On colonnaded structures the paint was confined to certain elements of the superstructure, from the level of the column capitals up through parts of the architrave, frieze, cornice, ceilings, and roof (37, 57, and front cover). Vivid red, blue, and sometimes green were the standard colors, and the use of gilding was not infrequent. The color was generally applied in the encaustic technique, whereby mineral pigments were mixed with molten wax, spread on in liquid form with a brush, and burned in with heated metal rods. The eaves tiles of roofs and the faces of simas, whether of marble or terracotta, were also usually painted with elaborate designs such as the bead-and-reel (58), egg-and-dart (37, inner band), Lesbian leaf (58), lotus-and-palmette (42), or meander pattern (42, 43). A given molding tended always to carry a specific design which could be either carved and painted (58) or simply painted. The interior walls, especially of houses, from the 5th century B.C. onwards, were also painted. The colors were black, white, yellow, but primarily red, often depicting masonry construction.

The most elaborate architectural decoration on public buildings was sculpture, which could appear in any or all of three locations and usually portrayed mythological scenes. One location was the frieze, where it was carved in relief on the metopes of the Doric order (32) or all along the continuous band of the Ionic (59, over inner columns). A second location was the triangular pediment at the gable end of a temple roof; here the figures were either in high relief or carved in the round. Finally, there were freestanding acroteria set on each end of the apex of the roof and on the four

59. The Hephaisteion, 5th century B.C.

corners; these generally represented either floral forms or additional mythological figures.

Many of the decorative elements of ancient buildings are lost, but the skills by which they were produced are listed in Plutarch's account, written in the 1st century after Christ, of a classical Athenian building program (*Life of Perikles*, 12):

> The materials to be used were stone, bronze, ivory, gold, ebony, and cypress wood; the arts which should elaborate and work up these materials were those of carpenter, molder, bronzesmith, stonecutter, dyer, worker in gold and ivory, painter, embroiderer, embosser, to say nothing of the forwarders and furnishers of the material, such as factors, sailors, and pilots by sea, and, by land, wagon makers, trainers of yoked beasts, and drivers. There were also rope makers, weavers, leather workers, road builders, and miners.

The passage clearly illustrates the range of materials and techniques in Athenian construction and shows that these buildings required the participation of large numbers of people with many varied skills. The temples, stoas, civic buildings, and private houses of the Agora remain as eloquent testimony of the industry and ingenuity of the Athenians throughout the ages.